AF677549

JUST THE FAX!

The World's All-Time Greatest FAX Cover Sheets

Wesley Eure & R. J. Jeffries

Illustrated by
Kersti O'Leary

KNIGHTSBRIDGE PUBLISHING COMPANY
New York

Copyright © 1990 by Wesley Eure and R. J. Jeffries

All rights reserved. Printed in the U.S.A.

Published in the United States by
Knightsbridge Publishing Company
255 East 49th Street
New York, New York 10017

ISBN: 1–877961–59–0

Library of Congress Cataloging-in-Publication Data
Eure, Wesley.
Just the Fax / Wesley Eure & R. J. Jeffries.—1st ed.
p. cm.
ISBN 1–877961–59–0: $9.95
1. Facsimile transmission—Caricatures and cartoons.
2. American wit and humor. Pictorial. I. Jeffries, R. J. II. Title.
NC1429.E8A4 1990
741.5'973—dc20 90-39030 CIP

10 9 8 7 6 5 4 3 2 1
First Edition

Contents

The FAX about Work

TRANSMITTAL COVER SHEET

JUST THE FAX!

RECEIVER:	SENDER:
Attention	Name
Company Name	Company
Phone Number	Phone Number
FAX Number	FAX Number

TOTAL PAGES ◯ **(with this page)**

© KNIGHTSBRIDGE PUBLISHING

GOOD MORNING!

TRANSMITTAL COVER SHEET

JUST THE FAX!

RECEIVER:	SENDER:
Attention	Name
Company Name	Company
Phone Number	Phone Number
FAX Number	FAX Number

TOTAL PAGES ◯ **(with this page)**

© KNIGHTSBRIDGE PUBLISHING

Occasion ____________________

Place ____________________

____

Time ____________________

R.S.V.P. ____________________

Attire ____________________

TRANSMITTAL COVER SHEET

JUST THE FAX!

RECEIVER:	SENDER:
Attention	Name
Company Name	Company
Phone Number	Phone Number
FAX Number	FAX Number

TOTAL PAGES ◯ **(with this page)**

© KNIGHTSBRIDGE PUBLISHING

Guess you're pretty popular

We couldn't get through until now!

JUST THE FAX!

TRANSMITTAL COVER SHEET

RECEIVER:	**SENDER:**
Attention	Name
Company Name	Company
Phone Number	Phone Number
FAX Number	FAX Number

TOTAL PAGES ◯ **(with this page)**

© KNIGHTSBRIDGE PUBLISHING

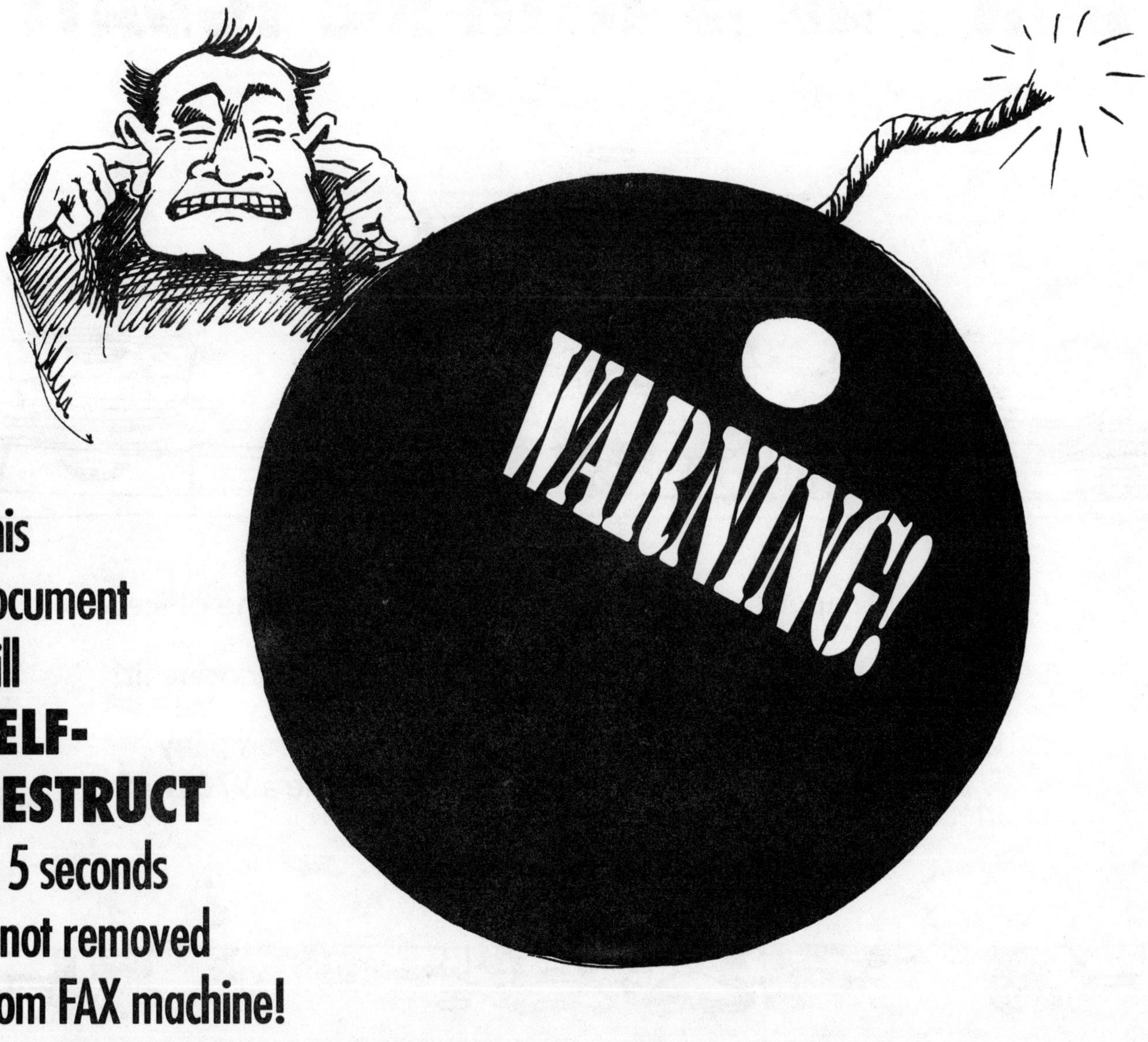

This document will **SELF-DESTRUCT** in 5 seconds if not removed from FAX machine!

TRANSMITTAL COVER SHEET

JUST THE FAX!

RECEIVER:	SENDER:
Attention	Name
Company Name	Company
Phone Number	Phone Number
FAX Number	FAX Number

TOTAL PAGES ◯ **(with this page)**

© KNIGHTSBRIDGE PUBLISHING

DON'T MESS WITH ME TODAY!

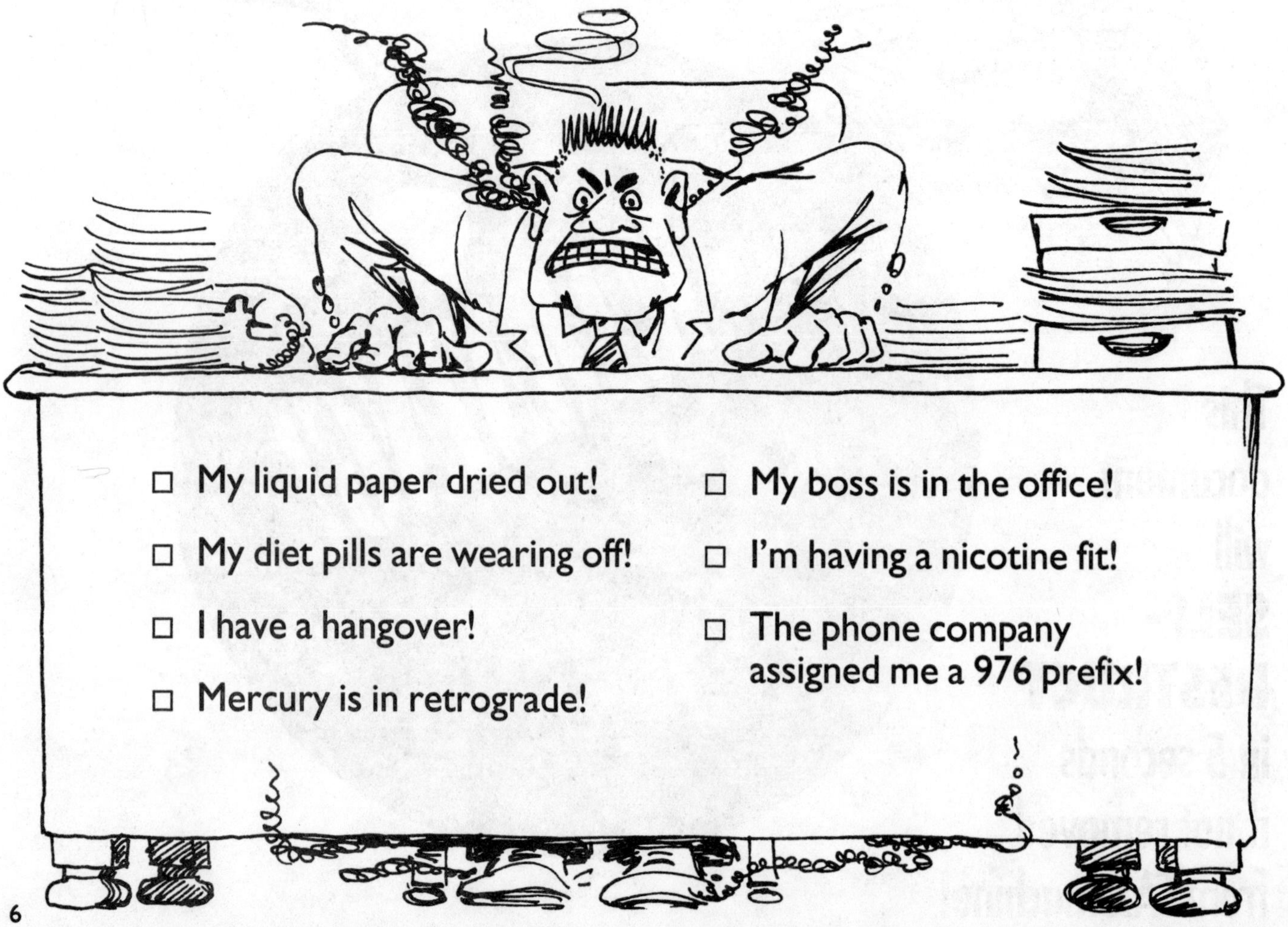

- ☐ My liquid paper dried out!
- ☐ My diet pills are wearing off!
- ☐ I have a hangover!
- ☐ Mercury is in retrograde!
- ☐ My boss is in the office!
- ☐ I'm having a nicotine fit!
- ☐ The phone company assigned me a 976 prefix!

TRANSMITTAL COVER SHEET

JUST THE FAX!

RECEIVER:	**SENDER:**
Attention	Name
Company Name	Company
Phone Number	Phone Number
FAX Number	FAX Number

TOTAL PAGES ◯ **(with this page)**

© KNIGHTSBRIDGE PUBLISHING

Congratulations

On Your Promotion!

TRANSMITTAL COVER SHEET

JUST THE FAX!

RECEIVER:	**SENDER:**
Attention	Name
Company Name	Company
Phone Number	Phone Number
FAX Number	FAX Number

TOTAL PAGES ◯ **(with this page)**

© KNIGHTSBRIDGE PUBLISHING

Congratulations

TRANSMITTAL COVER SHEET

JUST THE FAX!

RECEIVER:	**SENDER:**
Attention	Name
Company Name	Company
Phone Number	Phone Number
FAX Number	FAX Number

TOTAL PAGES ◯ **(with this page)**

© KNIGHTSBRIDGE PUBLISHING

Notice of

☐ **Meeting**

☐ **Conference Call**

With: ____________________

Date: ____________ Time: ____________

Re: ____________________

Place: ____________________

TRANSMITTAL COVER SHEET

JUST THE FAX!

RECEIVER:	SENDER:
Attention	Name
Company Name	Company
Phone Number	Phone Number
FAX Number	FAX Number

TOTAL PAGES ◯ **(with this page)**

© KNIGHTSBRIDGE PUBLISHING

Sorry But I Must Cancel

- ☐ Meeting
- ☐ Conference call

With ____________________________________

Date ____________________ Time ____________________

Place ____________________________________

Reason ____________________________________

TRANSMITTAL COVER SHEET

JUST THE FAX!

RECEIVER:	**SENDER:**
Attention	Name
Company Name	Company
Phone Number	Phone Number
FAX Number	FAX Number

TOTAL PAGES ◯ **(with this page)**

© KNIGHTSBRIDGE PUBLISHING

URGENT!

Please FAX or call at once!

Telephone number is: ____________________

FAX number is: ____________________

TRANSMITTAL COVER SHEET

JUST THE FAX!

RECEIVER:	**SENDER:**
Attention	Name
Company Name	Company
Phone Number	Phone Number
FAX Number	FAX Number

TOTAL PAGES ◯ **(with this page)**

© KNIGHTSBRIDGE PUBLISHING

TRANSMITTAL COVER SHEET

JUST THE FAX!	RECEIVER:	SENDER:
	Attention	Name
	Company Name	Company
	Phone Number	Phone Number
	FAX Number	FAX Number

TOTAL PAGES ◯ **(with this page)**

© KNIGHTSBRIDGE PUBLISHING

SORRY, BUT IT'S BEEN A ZOO AROUND HERE LATELY!

TRANSMITTAL COVER SHEET

JUST THE FAX!

RECEIVER:	SENDER:
Attention	Name
Company Name	Company
Phone Number	Phone Number
FAX Number	FAX Number

TOTAL PAGES ◯ **(with this page)**

© KNIGHTSBRIDGE PUBLISHING

PLEASE!

Don't go over my head

It only causes problems!

TRANSMITTAL COVER SHEET

JUST THE FAX!

RECEIVER:	**SENDER:**
Attention	Name
Company Name	Company
Phone Number	Phone Number
FAX Number	FAX Number

TOTAL PAGES ◯ **(with this page)**

KNIGHTSBRIDGE PUBLISHING

© KNIGHTSBRIDGE PUBLISHING

STOP GIVING ME THE RUN AROUND

TRANSMITTAL COVER SHEET

JUST THE FAX!

RECEIVER:	**SENDER:**
Attention	Name
Company Name	Company
Phone Number	Phone Number
FAX Number	FAX Number

TOTAL PAGES ◯ **(with this page)**

© KNIGHTSBRIDGE PUBLISHING

FAXTAG

Sorry we keep missing each other…

Have your machine call my machine!

TRANSMITTAL COVER SHEET

JUST THE FAX!

RECEIVER:	**SENDER:**
Attention	Name
Company Name	Company
Phone Number	Phone Number
FAX Number	FAX Number

TOTAL PAGES ◯ **(with this page)**

© KNIGHTSBRIDGE PUBLISHING

New Address: ______________________

New Telephone: ______________________

New FAX: ______________________

New Telex: ______________________

TRANSMITTAL COVER SHEET

JUST THE FAX!

RECEIVER:	SENDER:
Attention	Name
Company Name	Company
Phone Number	Phone Number
FAX Number	FAX Number

TOTAL PAGES ◯ **(with this page)**

© KNIGHTSBRIDGE PUBLISHING

TRANSMITTAL COVER SHEET

JUST THE FAX!

RECEIVER:	SENDER:
Attention	Name
Company Name	Company
Phone Number	Phone Number
FAX Number	FAX Number

TOTAL PAGES ◯ **(with this page)**

© KNIGHTSBRIDGE PUBLISHING

Quote of the Week

____

____

____

____

____

____

____

____

____

Said by: ______________ Heard by: ______________

Re: ______________________________

TRANSMITTAL COVER SHEET

JUST THE FAX!

RECEIVER:	**SENDER:**
Attention	Name
Company Name	Company
Phone Number	Phone Number
FAX Number	FAX Number

TOTAL PAGES ◯ **(with this page)**

© KNIGHTSBRIDGE PUBLISHING

LET'S DO LUNCH

Date: ______________________________

Time: ________________________

Restaurant: ______________

____

____

TRANSMITTAL COVER SHEET

JUST THE FAX!

RECEIVER:	SENDER:
Attention	Name
Company Name	Company
Phone Number	Phone Number
FAX Number	FAX Number

TOTAL PAGES ◯ **(with this page)**

© KNIGHTSBRIDGE PUBLISHING

WE OVERCHARGED YOU!

Credit Amount: $ ______________________________

______________________________ ______________________________

(signature) (title)

TRANSMITTAL COVER SHEET

JUST THE FAX!

RECEIVER:	**SENDER:**
Attention	Name
Company Name	Company
Phone Number	Phone Number
FAX Number	FAX Number

TOTAL PAGES ◯ **(with this page)**

© KNIGHTSBRIDGE PUBLISHING

We have looked all over for your check

Could it be...

☐ The check is in the mail?

☐ Your boss skipped to Mexico?

☐ You filed our invoice under chapter eleven?

☐ Other ______________________________

TRANSMITTAL COVER SHEET

JUST THE FAX!

RECEIVER:	SENDER:
Attention	Name
Company Name	Company
Phone Number	Phone Number
FAX Number	FAX Number

TOTAL PAGES ◯ **(with this page)**

© KNIGHTSBRIDGE PUBLISHING

Just a reminder...

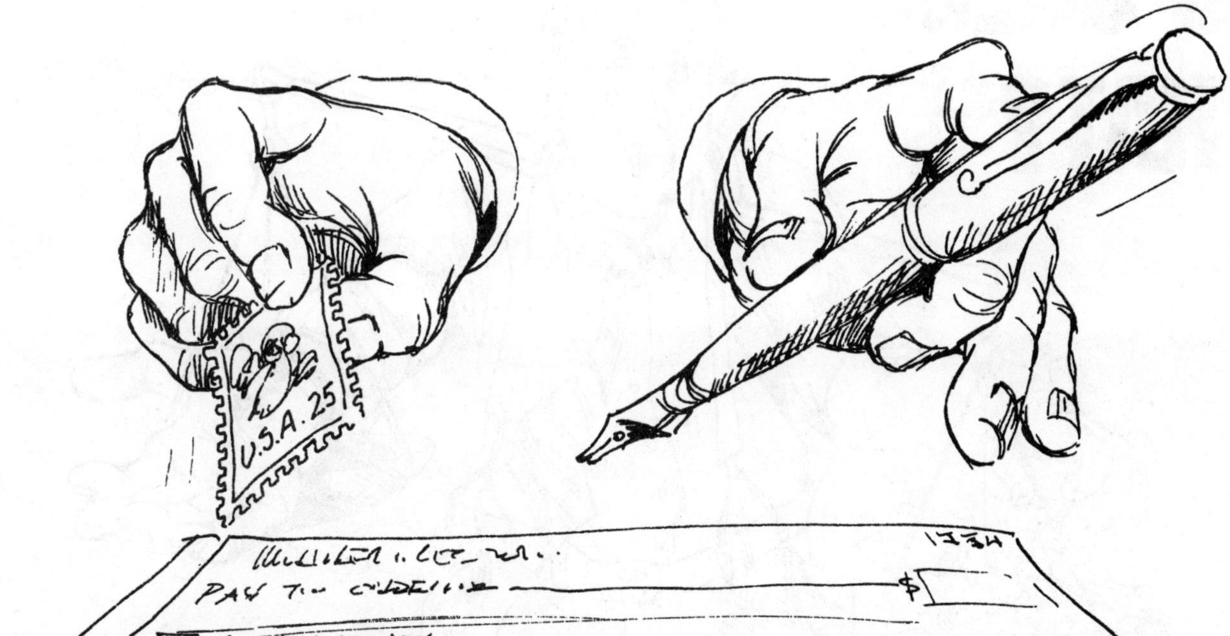

your payment is due now!

TRANSMITTAL COVER SHEET

JUST THE FAX!

RECEIVER:	SENDER:
Attention	Name
Company Name	Company
Phone Number	Phone Number
FAX Number	FAX Number

TOTAL PAGES ◯ **(with this page)**

© KNIGHTSBRIDGE PUBLISHING

THEY'RE CHARGING US MORE... SO GUESS WHAT?

As of ______________________ our prices will increase.

Old price: __________ New price: ______________

TRANSMITTAL COVER SHEET

JUST THE FAX!

RECEIVER:	**SENDER:**
Attention	Name
Company Name	Company
Phone Number	Phone Number
FAX Number	FAX Number

TOTAL PAGES ◯ **(with this page)**

© KNIGHTSBRIDGE PUBLISHING

Let's call it QUITS!

We're sorry, but we feel we can no longer do business with you...

Please:

- ☐ Pay outstanding invoices
- ☐ Return all of our materials
- ☐ Bill us
- ☐ Make good on any unfinished work
- ☐ Don't contact us anymore
- ☐ Contact ______________________________

Reason ______________________________

____

____

____

____

____

Signature ______________________________

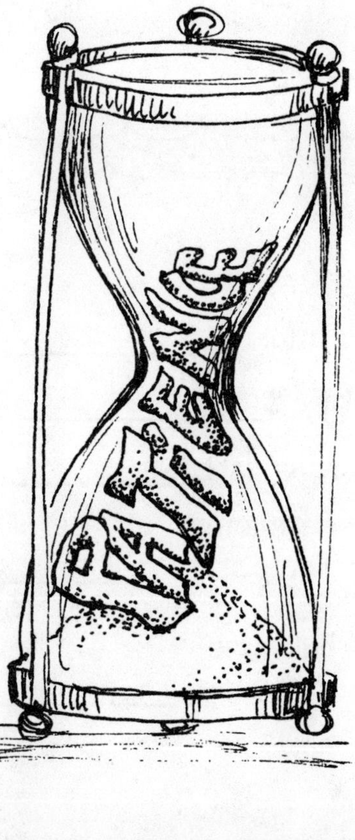

TRANSMITTAL COVER SHEET

JUST THE FAX!

RECEIVER:	SENDER:
Attention	Name
Company Name	Company
Phone Number	Phone Number
FAX Number	FAX Number

TOTAL PAGES ◯ **(with this page)**

© KNIGHTSBRIDGE PUBLISHING

HELP WANTED

Job title: ______________________

Salary: $__________ per ________

Starting date: ______________________

☐ Permanent

☐ Temporary ______________________ (ending date)

☐ Full-time

☐ Part-time ______________________ (days/hours)

Job description: ______________________

Job qualifications: ______________________

____

____

Please contact: ______________________

Resume required:

☐ Yes

☐ No

TRANSMITTAL COVER SHEET

JUST THE FAX!

RECEIVER:	SENDER:
Attention	Name
Company Name	Company
Phone Number	Phone Number
FAX Number	FAX Number

TOTAL PAGES ◯ **(with this page)**

© KNIGHTSBRIDGE PUBLISHING

Job Wanted

Type of job: ____________________

Salary: $__________ per __________

Availability date: ____________________

☐ Permanent

☐ Temporary ____________________

(ending date)

☐ Full-time

☐ Part-time ____________________

(days/hours)

Job description: ____________________

My qualifications: ____________________

____

____

My address: ____________________

Work #: __________ Home #: __________

TRANSMITTAL COVER SHEET

JUST THE FAX!

RECEIVER:	SENDER:
Attention	Name
Company Name	Company
Phone Number	Phone Number
FAX Number	FAX Number

TOTAL PAGES ◯ **(with this page)**

© KNIGHTSBRIDGE PUBLISHING

REQUEST FOR QUOTE

Please FAX me your quote re:

☐ The attached pages

☐ The info below

Quote needed by ______________________
(date/time)

Item # ______________________

Part # ______________________

Quantity ______________________

Price ______________________

Special instructions ______________________

____

____

____

____

TRANSMITTAL COVER SHEET

JUST THE FAX!

RECEIVER:	SENDER:
Attention	Name
Company Name	Company
Phone Number	Phone Number
FAX Number	FAX Number

TOTAL PAGES ◯ **(with this page)**

© KNIGHTSBRIDGE PUBLISHING

QUOTE *AS PER YOUR REQUEST*

The quote you requested for the following items:

Item #

____

Part #

____

Quantity

____

Price

____

Prices effective

____

(date)

____

(signature)

TRANSMITTAL COVER SHEET

JUST THE FAX!

RECEIVER:	SENDER:
Attention	Name
Company Name	Company
Phone Number	Phone Number
FAX Number	FAX Number

TOTAL PAGES ◯ **(with this page)**

© KNIGHTSBRIDGE PUBLISHING

FAX
AWARDS

TRANSMITTAL COVER SHEET

JUST THE FAX!

RECEIVER:	**SENDER:**
Attention	Name
Company Name	Company
Phone Number	Phone Number
FAX Number	FAX Number

TOTAL PAGES ◯ **(with this page)**

© KNIGHTSBRIDGE PUBLISHING

TRANSMITTAL COVER SHEET

JUST THE FAX!

RECEIVER:	**SENDER:**
Attention	Name
Company Name	Company
Phone Number	Phone Number
FAX Number	FAX Number

TOTAL PAGES ◯ **(with this page)**

© KNIGHTSBRIDGE PUBLISHING

Best Secretary Award

of the ❑ Day ❑ Month ❑ Year

Presented to: ______________________________

City: ____________________ Dept: ____________________

From: ____________________ Title: ____________________

TRANSMITTAL COVER SHEET

JUST THE FAX!

RECEIVER:	SENDER:
Attention	Name
Company Name	Company
Phone Number	Phone Number
FAX Number	FAX Number

TOTAL PAGES ◯ **(with this page)**

© KNIGHTSBRIDGE PUBLISHING

BEST SALESPERSON AWARD

of the

❑ Day ❑ Month ❑ Year

Presented to:

____

City: ________________________

Dept: ________________________

From: ________________________

Title: ________________________

TRANSMITTAL COVER SHEET

JUST THE FAX!

RECEIVER:	SENDER:
Attention	Name
Company Name	Company
Phone Number	Phone Number
FAX Number	FAX Number

TOTAL PAGES ◯ **(with this page)**

© KNIGHTSBRIDGE PUBLISHING

WORST

- ❑ tYping
- ❑ Speling
- ❑ Profffreading

AWARD

of the ❑ Day ❑ Month ❑ Year

Presented to: ______________________

City: ______________________

Dept: ______________________

From: ______________________

Title: ______________________

TRANSMITTAL COVER SHEET

JUST THE FAX!

RECEIVER:	SENDER:
Attention	Name
Company Name	Company
Phone Number	Phone Number
FAX Number	FAX Number

TOTAL PAGES ◯ **(with this page)**

© KNIGHTSBRIDGE PUBLISHING

Prettiest

❑ Salesman ❑ Boss ❑ Secretary

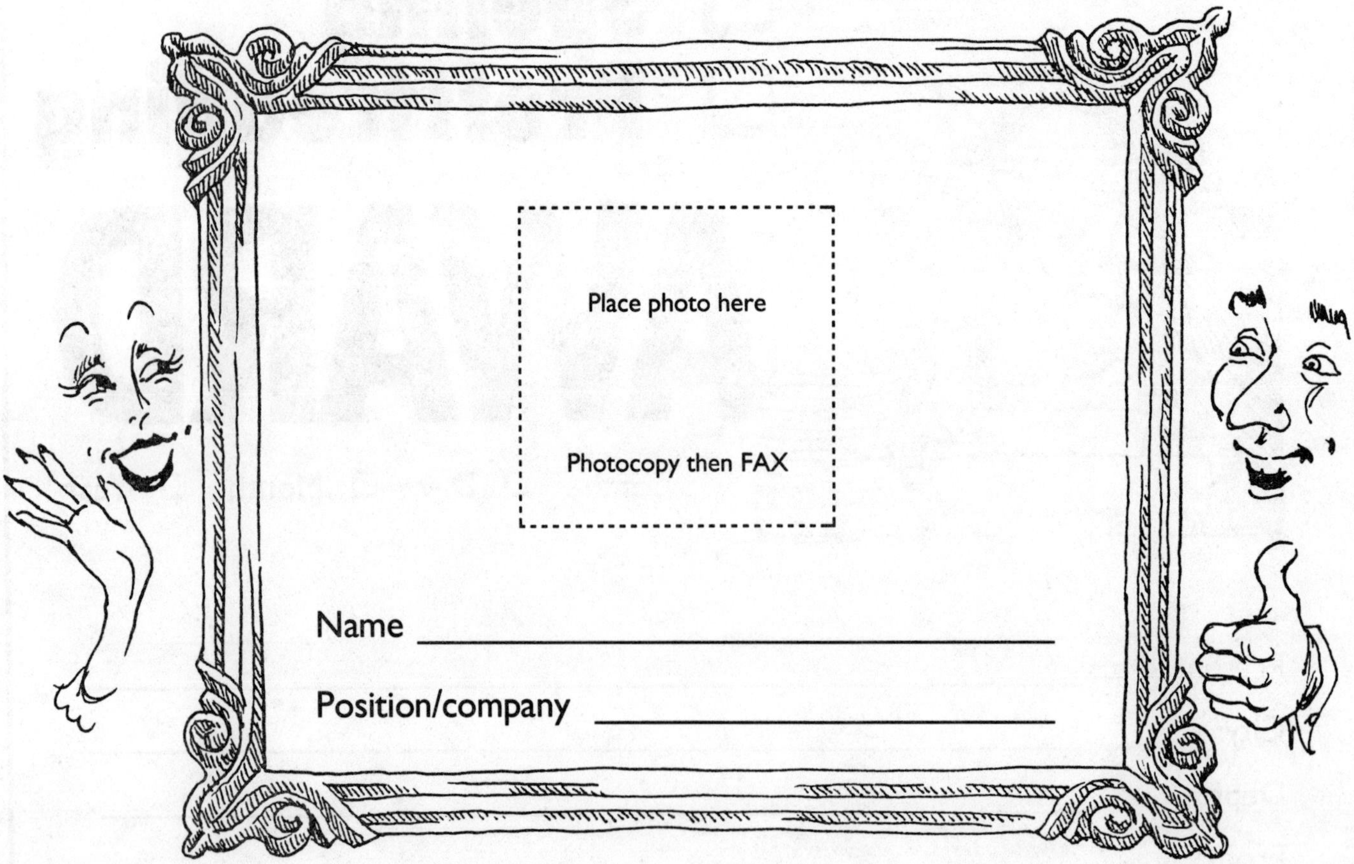

This contest is for the month of ____________.

Winners will receive a date with their chief executive officer (CEO). Remember, beauty is in the eye of the beholder!

JUST THE FAX!

TRANSMITTAL COVER SHEET

RECEIVER:	**SENDER:**
Attention	Name
Company Name	Company
Phone Number	Phone Number
FAX Number	FAX Number

TOTAL PAGES ◯ **(with this page)**

© KNIGHTSBRIDGE PUBLISHING

Ugliest

❑ Salesman ❑ Boss ❑ Secretary

Place photo here

Photocopy then FAX

Name ______________________________

Position/company ________________________

This contest is for the month of ____________________.

Winners will receive a genuine paper sack for their head. Remember, beauty is in the eye of the beholder!

TRANSMITTAL COVER SHEET

JUST THE FAX!

RECEIVER:	SENDER:
Attention	Name
Company Name	Company
Phone Number	Phone Number
FAX Number	FAX Number

TOTAL PAGES ( ) **(with this page)**

© KNIGHTSBRIDGE PUBLISHING

Best Rumor Monger Award

of the ❑ Day ❑ Month ❑ Year

Presented to: ______________________

City: ______________________

Dept: ______________________

From: ______________________

Title: ______________________

TRANSMITTAL COVER SHEET

JUST THE FAX!

RECEIVER:	SENDER:
Attention	Name
Company Name	Company
Phone Number	Phone Number
FAX Number	FAX Number

TOTAL PAGES ◯ **(with this page)**

© KNIGHTSBRIDGE PUBLISHING

I.O.U.

I ______________________ of ______________________

(name) (dept/company)

owe

____

(name)

A BIG FAVOR

To redeem, please fill out the request form below and re-FAX

____

____

____

TRANSMITTAL COVER SHEET

JUST THE FAX!

RECEIVER:	**SENDER:**
Attention	Name
Company Name	Company
Phone Number	Phone Number
FAX Number	FAX Number

TOTAL PAGES ◯ **(with this page)**

© KNIGHTSBRIDGE PUBLISHING

Oscar Predictions

Here's your chance to predict who will win an Oscar this year. Select one nominee in each of the following eight categories. Ballots must be received by ____________. Good luck!

____
Best picture

____
Best director

____
Best actor

____
Best actress

____
Best supporting actor

____
Best supporting actress

____
Best original song

____
Best foreign film

TRANSMITTAL COVER SHEET

JUST THE FAX!

RECEIVER:	**SENDER:**
Attention	Name
Company Name	Company
Phone Number	Phone Number
FAX Number	FAX Number

TOTAL PAGES ◯ **(with this page)**

© KNIGHTSBRIDGE PUBLISHING

The Emmy Awards

Here's your chance to predict who will win an Emmy this year. Select one nominee in each of the following categories.

Ballots must be received by ________________.

____

Best drama series

____

Best comedy series

____

Best actor/drama series

____

Best actress/drama series

____

Best actor/comedy series

____

Best actress/comedy series

____

Outstanding miniseries

TRANSMITTAL COVER SHEET

JUST THE FAX!

RECEIVER:	SENDER:
Attention	Name
Company Name	Company
Phone Number	Phone Number
FAX Number	FAX Number

TOTAL PAGES ◯ **(with this page)**

© KNIGHTSBRIDGE PUBLISHING

Grammy Award Predictions

Record of the year

Song of the year

Album of the year

Best new artist

Best country song

Best R & B Song

Here's your chance to predict who will win a Grammy this year. Select one nominee in each of the following six categories. Ballots must be received by ______________.

TRANSMITTAL COVER SHEET

JUST THE FAX!

RECEIVER:	SENDER:
Attention	Name
Company Name	Company
Phone Number	Phone Number
FAX Number	FAX Number

TOTAL PAGES ◯ **(with this page)**

© KNIGHTSBRIDGE PUBLISHING

Tony Award Predictions

Can you predict who will win Broadway's most coveted awards? Select one nominee in each of the following categories. Predictions must be received by ______ . Good luck!

Best play

Best musical

Best actor in a play

Best actress in a play

Best actor in a musical

Best actress in a musical

Best choreographer

Best director

Best playwright

Best revival

FAXinating You:

Love and Romance

TRANSMITTAL COVER SHEET

JUST THE FAX!

RECEIVER:	SENDER:
Attention	Name
Company Name	Company
Phone Number	Phone Number
FAX Number	FAX Number

TOTAL PAGES ◯ **(with this page)**

© KNIGHTSBRIDGE PUBLISHING

Since we have never met . . .

I'm sending you a recent photo!

TRANSMITTAL COVER SHEET

JUST THE FAX!

RECEIVER:	SENDER:
Attention	Name
Company Name	Company
Phone Number	Phone Number
FAX Number	FAX Number

TOTAL PAGES ◯ **(with this page)**

© KNIGHTSBRIDGE PUBLISHING

Since we have never met . . .

I'm sending you a recent photo!

TRANSMITTAL COVER SHEET

JUST THE FAX!

RECEIVER:	SENDER:
Attention	Name
Company Name	Company
Phone Number	Phone Number
FAX Number	FAX Number

TOTAL PAGES ◯ **(with this page)**

© KNIGHTSBRIDGE PUBLISHING

From a Secret Admirer

To ________________________

You are so:

- ☐ *Beautiful*
- ☐ *Handsome*
- ☐ *Nice*
- ☐ *Funny*
- ☐ *Warm*
- ☐ *Cuddly*
- ☐ *Caring*
- ☐ *Genuine*

☐ *Other* ________________________

____

Maybe one day our hearts will meet!

TRANSMITTAL COVER SHEET

JUST THE FAX!

RECEIVER:	SENDER:
Attention	Name
Company Name	Company
Phone Number	Phone Number
FAX Number	FAX Number

TOTAL PAGES ◯ **(with this page)**

© KNIGHTSBRIDGE PUBLISHING

How about a date?

Date: ______________________

Time: ______________________

Place: ______________________

____

Dress: ______________________

____

TRANSMITTAL COVER SHEET

JUST THE FAX!

RECEIVER:	SENDER:
Attention	Name
Company Name	Company
Phone Number	Phone Number
FAX Number	FAX Number

TOTAL PAGES ◯ **(with this page)**

© KNIGHTSBRIDGE PUBLISHING

My Dearest ________________

Will You Marry Me?

Love, ________________________

TRANSMITTAL COVER SHEET

JUST THE FAX!

RECEIVER:	SENDER:
Attention	Name
Company Name	Company
Phone Number	Phone Number
FAX Number	FAX Number

TOTAL PAGES ◯ **(with this page)**

© KNIGHTSBRIDGE PUBLISHING

Happy ____ Anniversary!
(number of years)

FAX for Fun: Food and Reservations

TRANSMITTAL COVER SHEET

JUST THE FAX!

RECEIVER:	SENDER:
Attention	Name
Company Name	Company
Phone Number	Phone Number
FAX Number	FAX Number

TOTAL PAGES ◯ (with this page)

© KNIGHTSBRIDGE PUBLISHING

Restaurant Take-out Order

Food

1. ______
2. ______
3. ______
4. ______
5. ______
6. ______
7. ______
8. ______
9. ______
10. ______

Beverages

1. ______
2. ______
3. ______
4. ______
5. ______
6. ______

Special Requests

______

______

______

We will pick up at:

______ A.M./P.M.
(time)

TRANSMITTAL COVER SHEET

JUST THE FAX!

RECEIVER:	SENDER:
Attention	Name
Company Name	Company
Phone Number	Phone Number
FAX Number	FAX Number

TOTAL PAGES ◯ (with this page)

© KNIGHTSBRIDGE PUBLISHING

Catering Bid Request

Please let me know what is available from your company within the specifications outlined below.

Date of affair: ______________________

Type of affair: ______________________

Number of people: ______________________

Budget: $ ______________________

Location: ______________________

Time: ______________________

Average age of guests: ______________________

Type of food preferred: ______________________

☐ Sit down dinner
☐ Cocktail party
☐ Buffet
☐ Hors d'oeuvres
☐ Full bar

☐ Beer & wine only
☐ Soft drinks
☐ Office party
☐ Restaurant party
☐ Home party

☐ Other ______________________

Name: ______________________

Tel. #: (   ) ______________________

FAX #: (   ) ______________________

TRANSMITTAL COVER SHEET

JUST THE FAX!

RECEIVER:	**SENDER:**
Attention	Name
Company Name	Company
Phone Number	Phone Number
FAX Number	FAX Number

TOTAL PAGES ◯ **(with this page)**

© KNIGHTSBRIDGE PUBLISHING

Recipe of the Week

TRANSMITTAL COVER SHEET

JUST THE FAX!

RECEIVER:	**SENDER:**
Attention	Name
Company Name	Company
Phone Number	Phone Number
FAX Number	FAX Number

TOTAL PAGES ◯ **(with this page)**

© KNIGHTSBRIDGE PUBLISHING

PLEASE!

❑ Confirm ❑ Cancel ❑ Make

confirmation # ______________________

my reservation for ______________________
(name)

on ______________ for ______________
(date/dates) (number of persons)

Amount of deposit (if any) $______________________

____
(signature)

Reservation Form

TRANSMITTAL COVER SHEET

JUST THE FAX!

RECEIVER:	**SENDER:**
Attention	Name
Company Name	Company
Phone Number	Phone Number
FAX Number	FAX Number

TOTAL PAGES ◯ **(with this page)**

© KNIGHTSBRIDGE PUBLISHING

Restaurant Reservation Form

Please make a reservation for ____________
at ____________ A.M./P.M. under the name
____ .

❏ Booth ❏ Table ❏ Cozy corner
❏ Private room ❏ Smoking ❏ Nonsmoking
❏ Handicapped seating for ____________

Please confirm:

Telephone #: (   ) ____________________ FAX #: (   ) ____________________

Additional information: ____

____

____

TRANSMITTAL COVER SHEET

JUST THE FAX!

RECEIVER:	SENDER:
Attention	Name
Company Name	Company
Phone Number	Phone Number
FAX Number	FAX Number

TOTAL PAGES ◯ **(with this page)**

© KNIGHTSBRIDGE PUBLISHING

Travel Plans

For: ____________________
(name)

From: ____________________

To: ____________________

Date: ____________________

Transportation: ____________________

Airline: ____________________

Flight #: ____________________

Departure: ____________________

Arrival: ____________________

Hotel: ____________________

Room #: ____________________

Hotel Phone #: __________ FAX #: __________

Special requests: ____________________

____

____

TRANSMITTAL COVER SHEET

JUST THE FAX!

RECEIVER:	SENDER:
Attention	Name
Company Name	Company
Phone Number	Phone Number
FAX Number	FAX Number

TOTAL PAGES ◯ **(with this page)**

© KNIGHTSBRIDGE PUBLISHING

Going on Vacation!

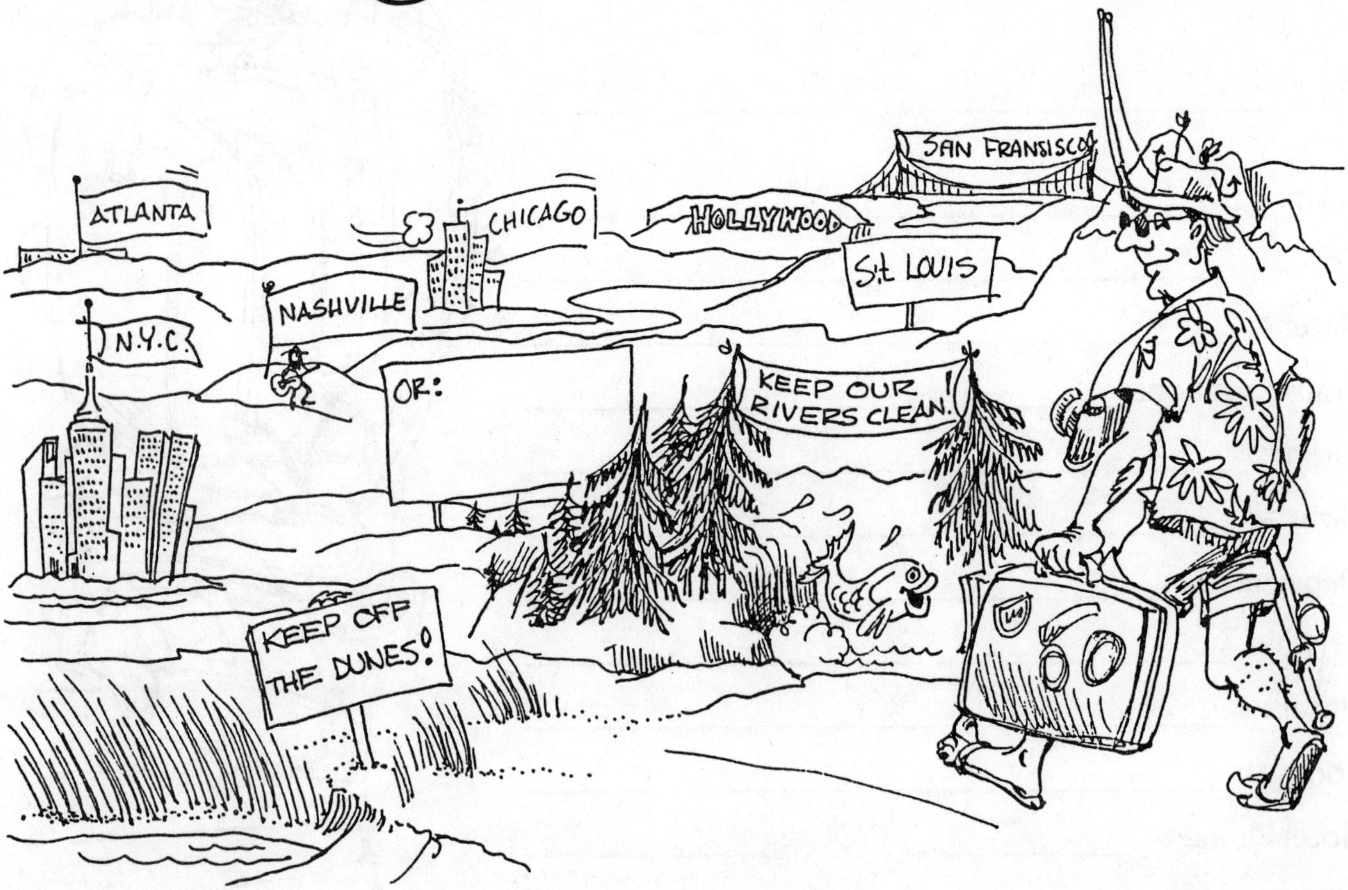

Leaving: ______________________ **Returning:** ______________________

TRANSMITTAL COVER SHEET

JUST THE FAX!

RECEIVER:	SENDER:
Attention	Name
Company Name	Company
Phone Number	Phone Number
FAX Number	FAX Number

TOTAL PAGES ◯ (with this page)

© KNIGHTSBRIDGE PUBLISHING

I'm off on a holiday!

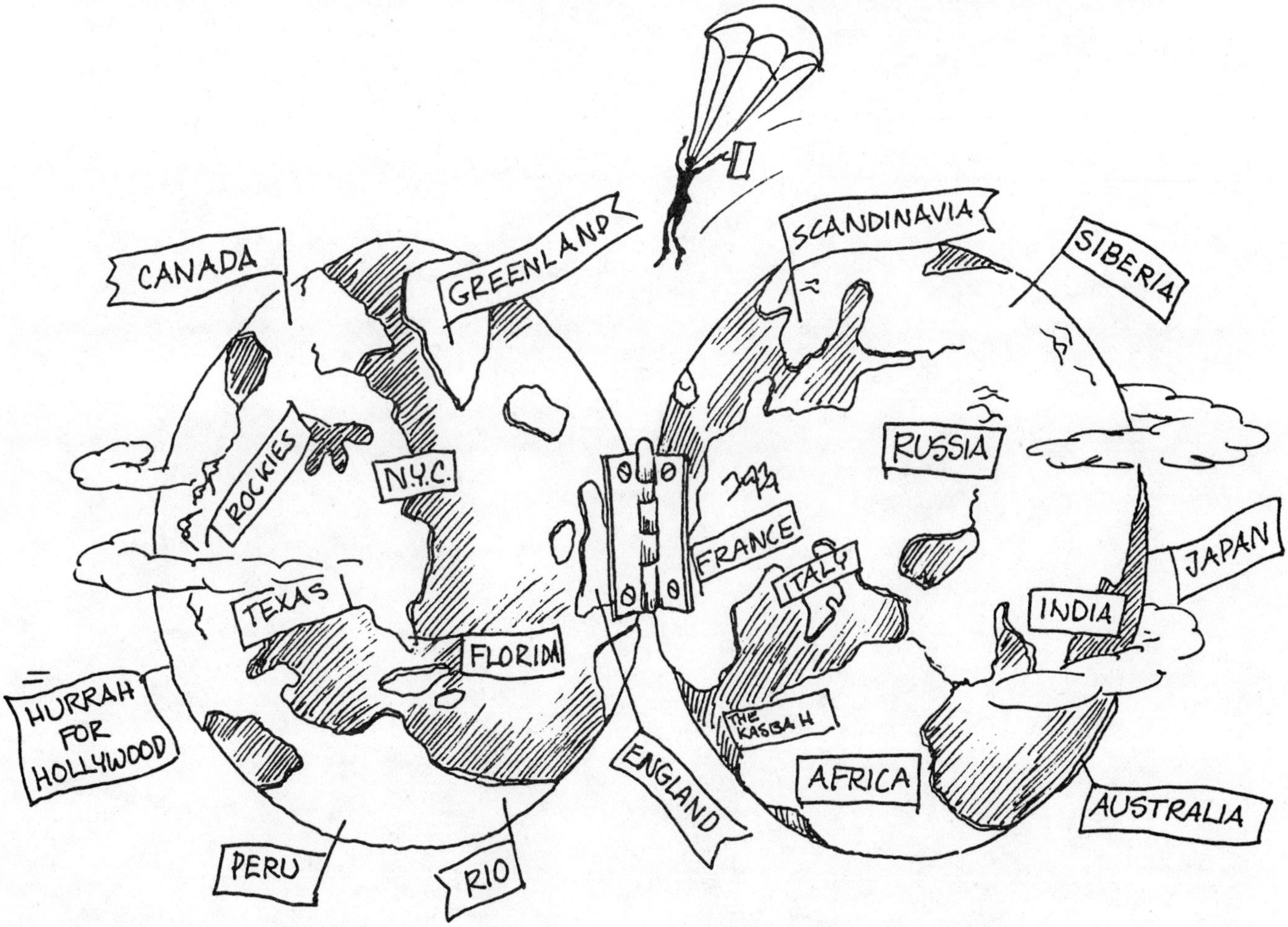

Leaving: ______________________ Returning: ______________________

Check out the map to see where I'm heading!

BEEP
BEEP
FAX
SALUTATIONS

TRANSMITTAL COVER SHEET

JUST THE FAX!

RECEIVER:	SENDER:
Attention	Name
Company Name	Company
Phone Number	Phone Number
FAX Number	FAX Number

TOTAL PAGES ◯ **(with this page)**

© KNIGHTSBRIDGE PUBLISHING

Happy Secretary's Day

Thanks for all your hard work!

TRANSMITTAL COVER SHEET

JUST THE FAX!

RECEIVER:	**SENDER:**
Attention	Name
Company Name	Company
Phone Number	Phone Number
FAX Number	FAX Number

TOTAL PAGES ◯ **(with this page)**

© KNIGHTSBRIDGE PUBLISHING

TRANSMITTAL COVER SHEET

JUST THE FAX!

RECEIVER:	SENDER:
Attention	Name
Company Name	Company
Phone Number	Phone Number
FAX Number	FAX Number

TOTAL PAGES ◯ **(with this page)**

© KNIGHTSBRIDGE PUBLISHING

TRANSMITTAL COVER SHEET

JUST THE FAX!

RECEIVER:	**SENDER:**
Attention	Name
Company Name	Company
Phone Number	Phone Number
FAX Number	FAX Number

TOTAL PAGES ◯ **(with this page)**

KNIGHTSBRIDGE PUBLISHING

© KNIGHTSBRIDGE PUBLISHING

TRANSMITTAL COVER SHEET

JUST THE FAX!

RECEIVER:	SENDER:
Attention	Name
Company Name	Company
Phone Number	Phone Number
FAX Number	FAX Number

TOTAL PAGES ◯ **(with this page)**

© KNIGHTSBRIDGE PUBLISHI

My Test was Positive!

I'm expecting a visit from the stork on

_____/_____/_____
(date)

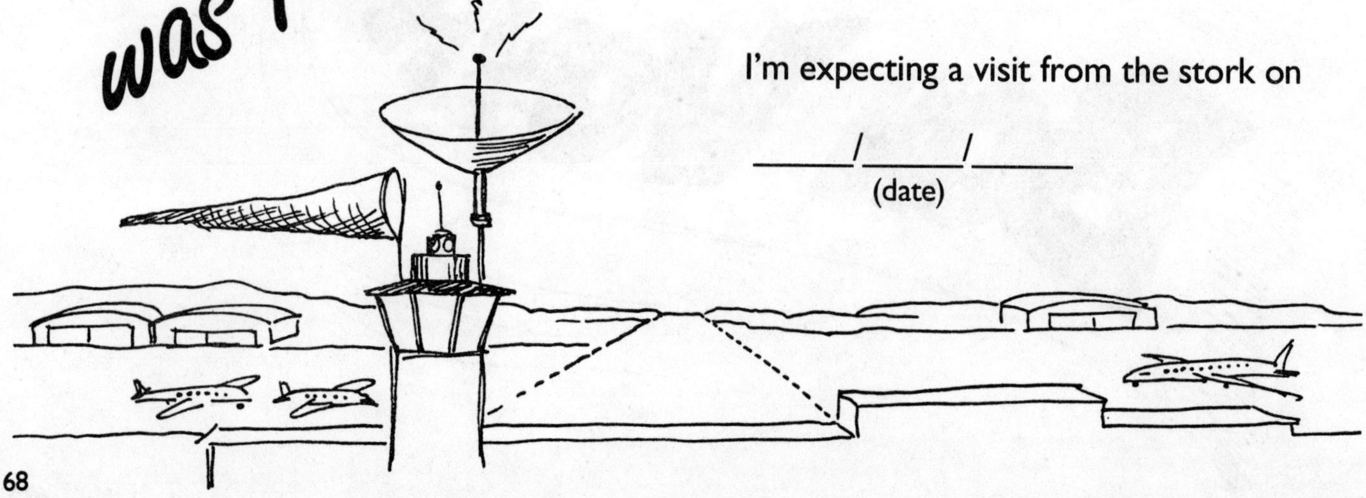

JUST THE FAX!

TRANSMITTAL COVER SHEET

RECEIVER:	**SENDER:**
Attention	Name
Company Name	Company
Phone Number	Phone Number
FAX Number	FAX Number

TOTAL PAGES ◯ **(with this page)**

© KNIGHTSBRIDGE PUBLISHING

Dear ______________________,

Will You Be My Valentine?

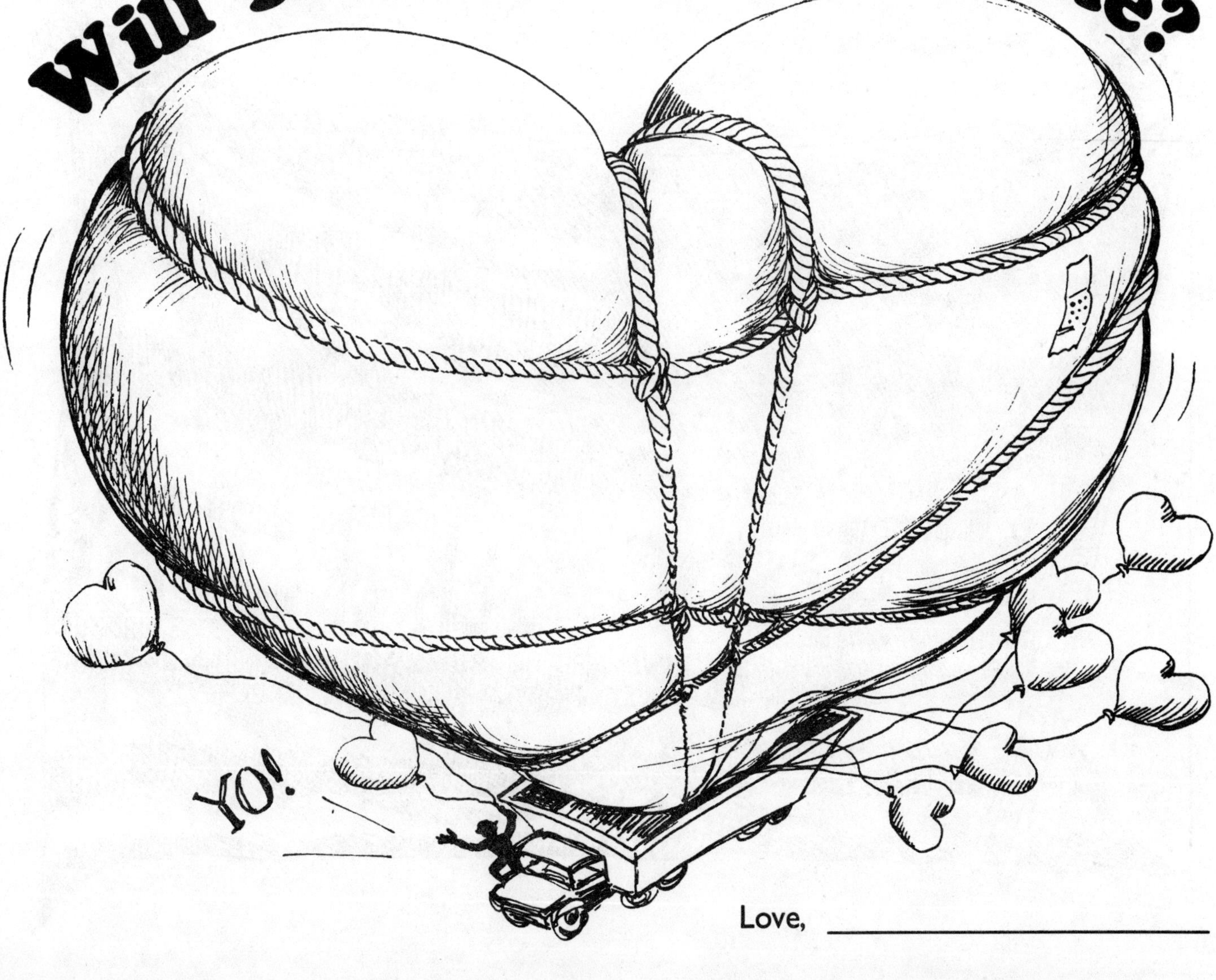

Love, ______________________

TRANSMITTAL COVER SHEET

JUST THE FAX!

RECEIVER:	**SENDER:**
Attention	Name
Company Name	Company
Phone Number	Phone Number
FAX Number	FAX Number

TOTAL PAGES ◯ **(with this page)**

© KNIGHTSBRIDGE PL

TRANSMITTAL COVER SHEET

JUST THE FAX!

RECEIVER:	SENDER:
Attention	Name
Company Name	Company
Phone Number	Phone Number
FAX Number	FAX Number

TOTAL PAGES ◯ **(with this page)**

© KNIGHTSBRIDGE PUBLISHING

TRANSMITTAL COVER SHEET

JUST THE FAX!

RECEIVER:	**SENDER:**
Attention	Name
Company Name	Company
Phone Number	Phone Number
FAX Number	FAX Number

TOTAL PAGES ◯ **(with this page)**

© KNIGHTSBRIDGE PUBLISHING

HAPPY HALLOWEEN!!!

TRANSMITTAL COVER SHEET

JUST THE FAX!

RECEIVER:	SENDER:
Attention	Name
Company Name	Company
Phone Number	Phone Number
FAX Number	FAX Number

TOTAL PAGES ◯ **(with this page)**

© KNIGHTSBRIDGE PUBLISHING

Happy Thanksgiving!

TRANSMITTAL COVER SHEET

JUST THE FAX!

RECEIVER:	SENDER:
Attention	Name
Company Name	Company
Phone Number	Phone Number
FAX Number	FAX Number

TOTAL PAGES ◯ **(with this page)**

© KNIGHTSBRIDGE PUBLISHING

Happy Chanukah

TRANSMITTAL COVER SHEET

JUST THE FAX!

RECEIVER:	**SENDER:**
Attention	Name
Company Name	Company
Phone Number	Phone Number
FAX Number	FAX Number

TOTAL PAGES ◯ **(with this page)**

© KNIGHTSBRIDGE PUBLISHING

We close on: ______________________ We reopen on: ______________________

TRANSMITTAL COVER SHEET

JUST THE FAX!

RECEIVER:	SENDER:
Attention	Name
Company Name	Company
Phone Number	Phone Number
FAX Number	FAX Number

TOTAL PAGES ( ) **(with this page)**

© KNIGHTSBRIDGE PUBLISHING

Happy New Year!

Auto FAX

TRANSMITTAL COVER SHEET

JUST THE FAX!

RECEIVER:	**SENDER:**
Attention	Name
Company Name	Company
Phone Number	Phone Number
FAX Number	FAX Number

TOTAL PAGES ◯ **(with this page)**

© KNIGHTSBRIDGE PUBLISHING

On your way home, please pick up:

- ☐ Personal items: ____________________
- ☐ Grocery items: ____________________
- ☐ Dry cleaning: ____________________
- ☐ Kids: ____________________
- ☐ Other: ____________________

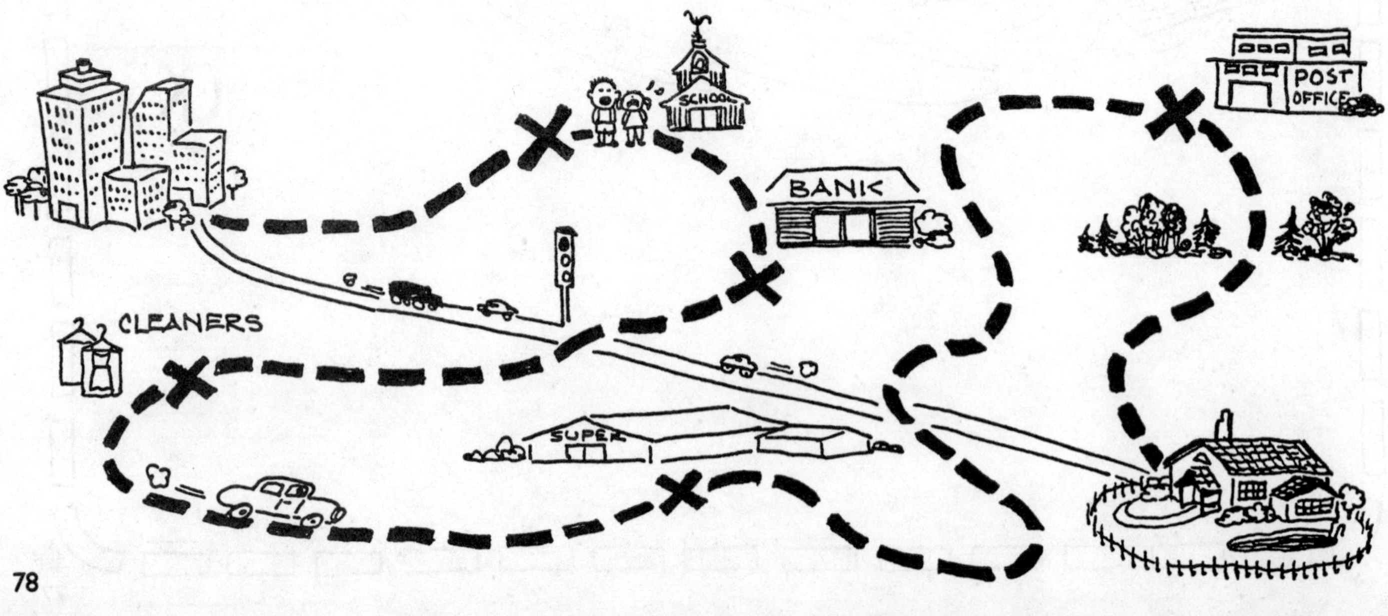

TRANSMITTAL COVER SHEET

JUST THE FAX!

RECEIVER:	SENDER:
Attention	Name
Company Name	Company
Phone Number	Phone Number
FAX Number	FAX Number

TOTAL PAGES ◯ **(with this page)**

© KNIGHTSBRIDGE PUBLISHING

On your way to the office, please pick up:

- ☐ Documents at: ____________________
- ☐ Office supplies at: ____________________
- ☐ Mail/parcel at: ____________________
- ☐ Coffee/pastries: ____________________
- ☐ Coworker at: ____________________
- ☐ Client (Mr./Ms.): ____________________ at: ____________________
- ☐ Other: ____________________

TRANSMITTAL COVER SHEET

JUST THE FAX!

RECEIVER:	SENDER:
Attention	Name
Company Name	Company
Phone Number	Phone Number
FAX Number	FAX Number

TOTAL PAGES ◯ **(with this page)**

© KNIGHTSBRIDGE PUBLISHING

I'm running late because:

- ☐ I have a flat tire
- ☐ I'm stuck in traffic
- ☐ I have engine trouble
- ☐ I didn't follow directions
- ☐ I followed your directions
- ☐ Traffic snarled
- ☐ The kids are screaming
- ☐ The dog needed a hydrant
- ☐ I have to tinkle
- ☐ My air conditioner is on the fritz
- ☐ I just passed a
  - ☐ blonde
  - ☐ redhead
  - ☐ brunette

  in a convertible and I'm following
  - ☐ him
  - ☐ her
- ☐ My
  - ☐ mother-in-law
  - ☐ husband
  - ☐ wife
  - ☐ kid
  - ☐ other ___________

  is bugging me
- ☐ I have a premonition
- ☐ I rear ended a blonde
- ☐ I got shot by a radar gun
- ☐ I threw my rod

TRANSMITTAL COVER SHEET

JUST THE FAX!

RECEIVER:	SENDER:
Attention	Name
Company Name	Company
Phone Number	Phone Number
FAX Number	FAX Number

TOTAL PAGES ◯ **(with this page)**

© KNIGHTSBRIDGE PUBLISHING

U-Turn

We know you are on your way, but the meeting has been:

☐ Cancelled

☐ Changed to: ___/___/___ at ________ A.M./P.M.
(date) (time)

☐ Set for new location:

____

Directions:

____

____

(Signature) (Title)

TRANSMITTAL COVER SHEET

JUST THE FAX!

RECEIVER:	SENDER:
Attention	Name
Company Name	Company
Phone Number	Phone Number
FAX Number	FAX Number

TOTAL PAGES ◯ **(with this page)**

© KNIGHTSBRIDGE PUBLISHING

The telephone number you requested for:

____________________ **is (   )**____________________

(name) **(number)**

TRANSMITTAL COVER SHEET

JUST THE FAX!

RECEIVER:	**SENDER:**
Attention	Name
Company Name	Company
Phone Number	Phone Number
FAX Number	FAX Number

TOTAL PAGES ◯ **(with this page)**

KNIGHTSBRIDGE PUBLISHING

© KNIGHTSBRIDGE PUBLISHING

Per your request, the nearest ____________________

is (name of establishment) ____________________,

at (address) ____________________.

Telephone: ____________________

Directions: ____________________

____

____

Location Scout

FAX OFF!

TRANSMITTAL COVER SHEET

JUST THE FAX!

RECEIVER:	**SENDER:**
Attention	Name
Company Name	Company
Phone Number	Phone Number
FAX Number	FAX Number

TOTAL PAGES ◯ **(with this page)**

© KNIGHTSBRIDGE PUBLISHING

Take this job and shove it!

- ❑ Where the sun don't shine!
- ❑ Where the sun don't shine!!
- ❑ Where the sun don't shine!!!
- ❑ Where the sun don't shine!!!!

TRANSMITTAL COVER SHEET

JUST THE FAX!

RECEIVER:	**SENDER:**
Attention	Name
Company Name	Company
Phone Number	Phone Number
FAX Number	FAX Number

TOTAL PAGES ( ) **(with this page)**

© KNIGHTSBRIDGE PUBLISHING

Why DID YOU DO THAT?

Just state the facts!

TRANSMITTAL COVER SHEET

JUST THE FAX!

RECEIVER:	SENDER:
Attention	Name
Company Name	Company
Phone Number	Phone Number
FAX Number	FAX Number

TOTAL PAGES ◯ **(with this page)**

KNIGHTSBRIDGE PUBLISHING

© KNIGHTSBRIDGE PUBLISHING

GET A LIFE !

TRANSMITTAL COVER SHEET

JUST THE FAX!

RECEIVER:	SENDER:
Attention	Name
Company Name	Company
Phone Number	Phone Number
FAX Number	FAX Number

TOTAL PAGES ◯ **(with this page)**

© KNIGHTSBRIDGE PUBLISHING

IF YOU HAVE SOMETHING TO SAY

SAY IT TO MY FACE!

Re: ______________________________

TRANSMITTAL COVER SHEET

JUST THE FAX!

RECEIVER:	**SENDER:**
Attention	Name
Company Name	Company
Phone Number	Phone Number
FAX Number	FAX Number

TOTAL PAGES ( ) **(with this page)**

© KNIGHTSBRIDGE PUBLISHING

I'm mad as hell, and I'm not going to take it anymore!

I need:
- ☐ a promotion!
- ☐ a vacation!
- ☐ some sick leave!
- ☐ a maternity leave!
- ☐ a rest!
- ☐ a new boss!
- ☐ some respect!
- ☐ tranquilizers!
- ☐ a cocktail!
- ☐ a Godfather!
- ☐ other _______

____

____

____

____

TRANSMITTAL COVER SHEET

JUST THE FAX!

RECEIVER:	SENDER:
Attention	Name
Company Name	Company
Phone Number	Phone Number
FAX Number	FAX Number

TOTAL PAGES ◯ **(with this page)**

© KNIGHTSBRIDGE PUBLISHING

I'm mad as hell, and I'm not going to take it anymore!

You need:

- ☐ a swift kick in the rear
- ☐ a sabbatical
- ☐ a psychiatrist
- ☐ an attorney
- ☐ a sex therapist
- ☐ a new line of work
- ☐ a divorce

- ☐ a good ____________ (fill in the blank)
- ☐ a lobotomy
- ☐ an attitude adjustment
- ☐ to get laid
- ☐ other ____________

____

TRANSMITTAL COVER SHEET

JUST THE FAX!

RECEIVER:	SENDER:
Attention	Name
Company Name	Company
Phone Number	Phone Number
FAX Number	FAX Number

TOTAL PAGES ◯ **(with this page)**

© KNIGHTSBRIDGE PUBLISHING

If I were you . . . I would do some serious soul searching!

FAX Attack: A Grab Bag

TRANSMITTAL COVER SHEET

JUST THE FAX!

RECEIVER:	SENDER:
Attention	Name
Company Name	Company
Phone Number	Phone Number
FAX Number	FAX Number

TOTAL PAGES ◯ **(with this page)**

© KNIGHTSBRIDGE PUBLISHING

Our ____________________ team
(sport)

the

____
(name)

was victorious

____
(date)

over the

____
(opponent's team)

The score was _______ to _______!

WE WON!

TRANSMITTAL COVER SHEET

JUST THE FAX!

RECEIVER:	SENDER:
Attention	Name
Company Name	Company
Phone Number	Phone Number
FAX Number	FAX Number

TOTAL PAGES ◯ **(with this page)**

KNIGHTSBRIDGE PUBLISHING

© KNIGHTSBRIDGE PUBLISHING

We Lost

Our ____________________ team
(sport)

the

____
(name)

was beaten

____
(date)

by the

______________________________.
(opponent's team)

The score was ______ to ______!

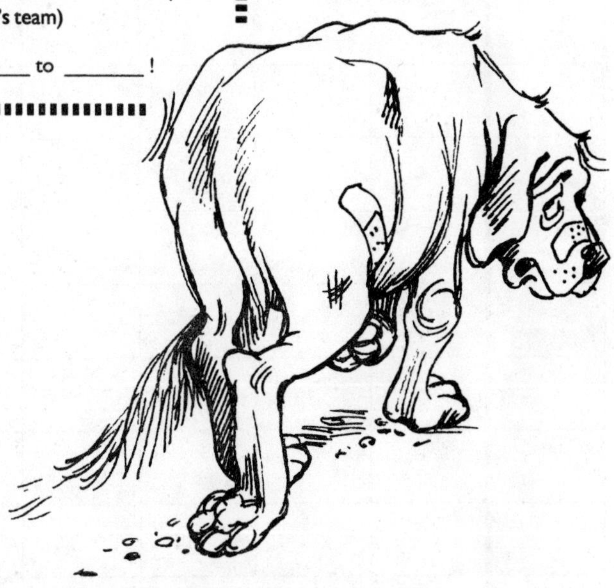

JUST THE FAX!

TRANSMITTAL COVER SHEET

RECEIVER:	SENDER:
Attention	Name
Company Name	Company
Phone Number	Phone Number
FAX Number	FAX Number

TOTAL PAGES ◯ **(with this page)**

© KNIGHTSBRIDGE PUBLISHING

SPORTS POOL

Number small slips of paper "0–9." Place them in a container and draw them one at a time, placing each number at the top of the form, from left to right. Follow the same procedure for the left side of the form. Draw the two teams' names and place one on the top of the paper and the other along the left side. There are 100 boxes. Everyone chooses one or more boxes and places their name inside. The winner is the person who comes the closest to the final score. Note: You only count the last numbers of the score. If the score is 21 to 30, the winner has the 0 for the winning team and the 1 for the losing team. Good luck!

	Team Name										
Team Name	Numbers										

TRANSMITTAL COVER SHEET

JUST THE FAX!

RECEIVER:	**SENDER:**
Attention	Name
Company Name	Company
Phone Number	Phone Number
FAX Number	FAX Number

TOTAL PAGES ◯ **(with this page)**

© KNIGHTSBRIDGE PUBLISHING

It's a

☐ Boy

☐ Girl

Name ______________________________

Date __________ Time ________ Weight __________ lbs. __________ oz.

TRANSMITTAL COVER SHEET

JUST THE FAX!

RECEIVER:	SENDER:
Attention	Name
Company Name	Company
Phone Number	Phone Number
FAX Number	FAX Number

TOTAL PAGES ◯ **(with this page)**

© KNIGHTSBRIDGE PUBLISHING

FAX DIET CLUB
JOIN TODAY!

Do you need a little support to
shed those unwanted pounds?
Then sign up for our FREE DIET CLUB!

It's simple

Just fax me the information requested below and I will then fax you a complete list of members.

On ___________ , we will begin our THREE MONTH quest for fitness!
(date)

At the end of each week, you will fax me your current weight and an updated list of everyone's progress will be faxed back.

Important

Anytime during the next few months that you feel a binge coming on, just get in touch with any of our members for support!

Together we can do it!

Just fax me your:

Name ____________________ Company ____________________ Phone # __________

FAX #________ Weight _____ Goal (pounds you wish to lose) _____

TRANSMITTAL COVER SHEET

JUST THE FAX!

RECEIVER:	SENDER:
Attention	Name
Company Name	Company
Phone Number	Phone Number
FAX Number	FAX Number

TOTAL PAGES ◯ **(with this page)**

© KNIGHTSBRIDGE PUBLISHING

FAX DIET CLUB
WEEKLY REPORT

Name					
FAX #					
Goal					
Lbs. lost Week #					
1					
2					
3					
4					
5					
6					
7					
8					
9					
10					
11					

TRANSMITTAL COVER SHEET

JUST THE FAX!

RECEIVER:	**SENDER:**
Attention	Name
Company Name	Company
Phone Number	Phone Number
FAX Number	FAX Number

TOTAL PAGES ◯ **(with this page)**

© KNIGHTSBRIDGE PUBLISHING

FAX QUIT-SMOKING CLUB

JOIN TODAY!

Do you want to stop smoking?
Together we can do it!

For the next 5 days beginning Monday ______________________,
(date)
a list of members and their progress will be faxed to you in the morning. Between 3:00 P.M. and 5:00 P.M., fax me your report on how well you have done for the day.

IMPORTANT

If ever the urge to smoke becomes too strong,

— STOP —

then call any of the other members.

If interested, fax me the info requested below:

Name ______________ Company ______________

Phone # ______________ FAX # ______________

TRANSMITTAL COVER SHEET

JUST THE FAX!

RECEIVER:	SENDER:
Attention	Name
Company Name	Company
Phone Number	Phone Number
FAX Number	FAX Number

TOTAL PAGES ◯ **(with this page)**

© KNIGHTSBRIDGE PUBLISHING

FAX QUIT-SMOKING CLUB
DAILY REPORT

Member				
FAX				
Phone				
Number of cigarettes smoked today				
Monday				
Tuesday				
Wednesday				
Thursday				
Friday				

TRANSMITTAL COVER SHEET

JUST THE FAX!

RECEIVER:	SENDER:
Attention	Name
Company Name	Company
Phone Number	Phone Number
FAX Number	FAX Number

TOTAL PAGES ◯ **(with this page)**

© KNIGHTSBRIDGE PUBLISHING

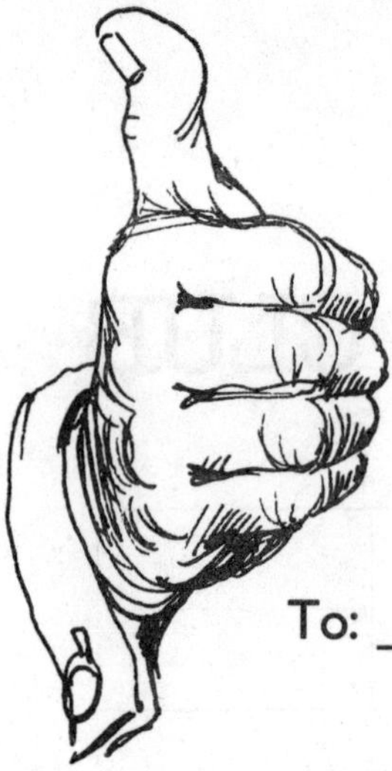

Please cast my vote!

To: ______________________________
(elected official)

From: ______________________________
(your name)

Address: ______________________________

____

Phone: ______________________________

FAX: ______________________________

Re: ______________________________
(bill/issue)

Please vote:

☐ Yes

☐ No

Reason: ______________________________

____

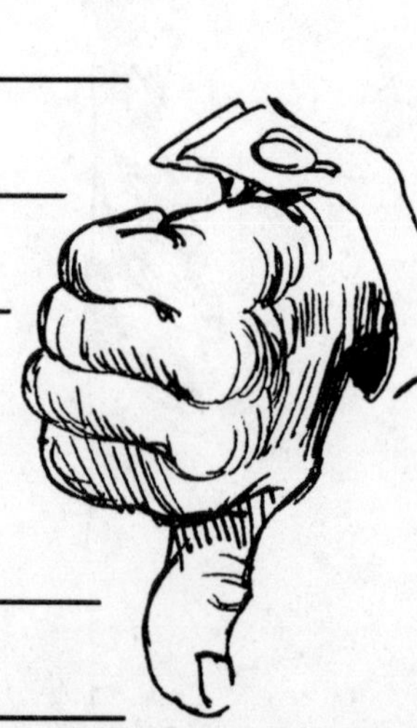

TRANSMITTAL COVER SHEET

JUST THE FAX!

RECEIVER:	**SENDER:**
Attention	Name
Company Name	Company
Phone Number	Phone Number
FAX Number	FAX Number

TOTAL PAGES ◯ **(with this page)**

© KNIGHTSBRIDGE PUBLISHING

I'm sorry!

Can't we put the pieces back together again?

TRANSMITTAL COVER SHEET

JUST THE FAX!

RECEIVER:	SENDER:
Attention	Name
Company Name	Company
Phone Number	Phone Number
FAX Number	FAX Number

TOTAL PAGES ◯ **(with this page)**

© KNIGHTSBRIDGE PUBLISHING

Best joke of the month contest

Please FAX us the funniest joke you have heard this month. Just fill in the form and send it to us by ____________ . The winner will receive this fabulous one-of-a-kind prize:

TRANSMITTAL COVER SHEET

JUST THE FAX!

RECEIVER:	**SENDER:**
Attention	Name
Company Name	Company
Phone Number	Phone Number
FAX Number	FAX Number

TOTAL PAGES ◯ **(with this page)**

© KNIGHTSBRIDGE PUBLISHING

Please Play

Song ____________________

Performed by ____________________

Requested by ____________________

Dedicated to ____________________

Special message ____________________

____

TRANSMITTAL COVER SHEET

JUST THE FAX!

RECEIVER:	**SENDER:**
Attention	Name
Company Name	Company
Phone Number	Phone Number
FAX Number	FAX Number

TOTAL PAGES ◯ **(with this page)**

© KNIGHTSBRIDGE PUBLISHING

Want Ads

For:

❑ Sale ❑ Lease ❑ Rent
❑ Trade ❑ Free

Description:

TRANSMITTAL COVER SHEET

JUST THE FAX!

RECEIVER:	**SENDER:**
Attention	Name
Knightsbridge Publishing Company Name	Company
Phone Number **(213) 274-3644**	Phone Number
FAX Number	FAX Number

TOTAL PAGES ◯ **(with this page)**

© KNIGHTSBRIDGE PUBLISHING

Dear friends,

Share the “FAX” with everyone!

Please FAX us any unusual experiences you have had using the *Just the FAX* cover sheets!

Did she say “yes” when you sent the marriage proposal?

Did the smoking club clear the air?

Did the recipients of the awards get angry or laugh?

We would also love to hear of any new *Just the FAX* cover sheets you want to see in our next book. (All faxed suggestions become the property of Knightsbridge Publishing and you give us complete permission to publish them. We can't pay you, but we will give you a mention in our forthcoming book!)

Thanks for joining our growing FAX family!

Wesley & R. J.